edZOOcation™ presents:

by Sara Karnoscak

Wildlife Tree
edZOOcation

SUPPORTED 2 READER

Dedication:

Widmet sich das buch Connor

–S.K.

For Noah Paz, the amature zookeeper.

–T.S.

Copyright © 2023 Wildlife Tree, LLC. All rights reserved.

Author: Sara Karnoscak

Designer: Tiffany Swicegood

Editor: Tess Riley

Photo Credits:

AdobeStock.com

Pixabay.com

Pexels.com

ISBN: 979-8-9886164-4-3

This book meets **Common Core** and **Next Generation Science** Standards.

Table of Contents

4   A Bat's Body

8   Walking, Pooping, and Pollinating

10   Vampires and Other Bats

14   Bat Colonies

16   Hear That?

17   Where Bats Hang Out

21   Sleeping and Eating

24   A Night in the Life of a Bat

26   The Food Web and Dangers

28   Comparing Hands

30   Glossary

32   Funny Flyers

A Bat's Body

Thumbs for grabbing and climbing.

Long fingers that support the wings.

A bat's wing is a giant hand! Webbing between the bat's long fingers makes the wing.

Big ears for echolocating.

Arms that form the tops of the wings.

Eyes that see best in dim light.

Echolocation: *To find out where things are by listening to an echo.*

Flying Solo

Mammal: Warm-blooded animals with hair or fur.

Bats are the only **mammals** that can fly. Some can fly as fast as 100 miles per hour! That's almost twice as fast as a cheetah can run.

Bats can swoop and change direction fast. This helps them catch bugs. It also helps them avoid flying into things.

Bats Helping People

Clot: *Blood that hardens into a clump.*

Vampire bat spit keeps blood from **clotting**. Scientists study them to learn how to keep people's blood from clotting.

Scientists also study echolocation to help people. People who can't see can learn echolocation to help them.

Can Bats Walk?

Bats are built to fly. Most bats can't really walk. But a few bats can walk.

Since their fingers make up their wing, they can only walk on their wrist and thumb. They use their strong wings to pull their body forward.

Some bats can even swim! They will swim to land if they fall in water.

Bats in the Garden

Guano: *A build-up of poop from bats or seabirds.*

Bat poop looks a lot like mouse poop. Lots of bat poop together is called **guano**. Bat guano is good for gardens. It can help plants grow.

Bats also help plants grow by spreading **pollen**. Bats are some of Earth's greatest **pollinators**.

Without bats, we wouldn't have many of the foods we love!

Pollen: *Dust that helps plants form seeds.*

Pollinator: *An animal that moves pollen from one plant to another.*

The *Bat*ting Line-Up

There are more than 1,300 species of bats. Microbats are small. As small as six inches. Megabats are bigger. As big as six feet.

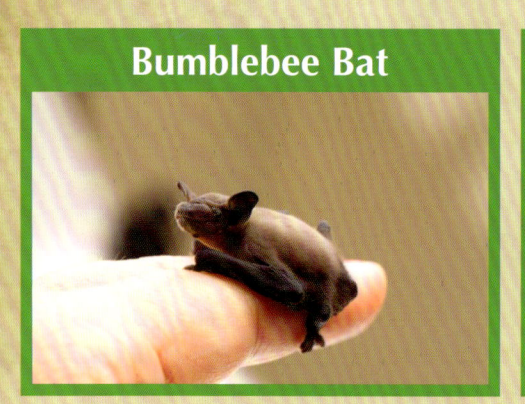

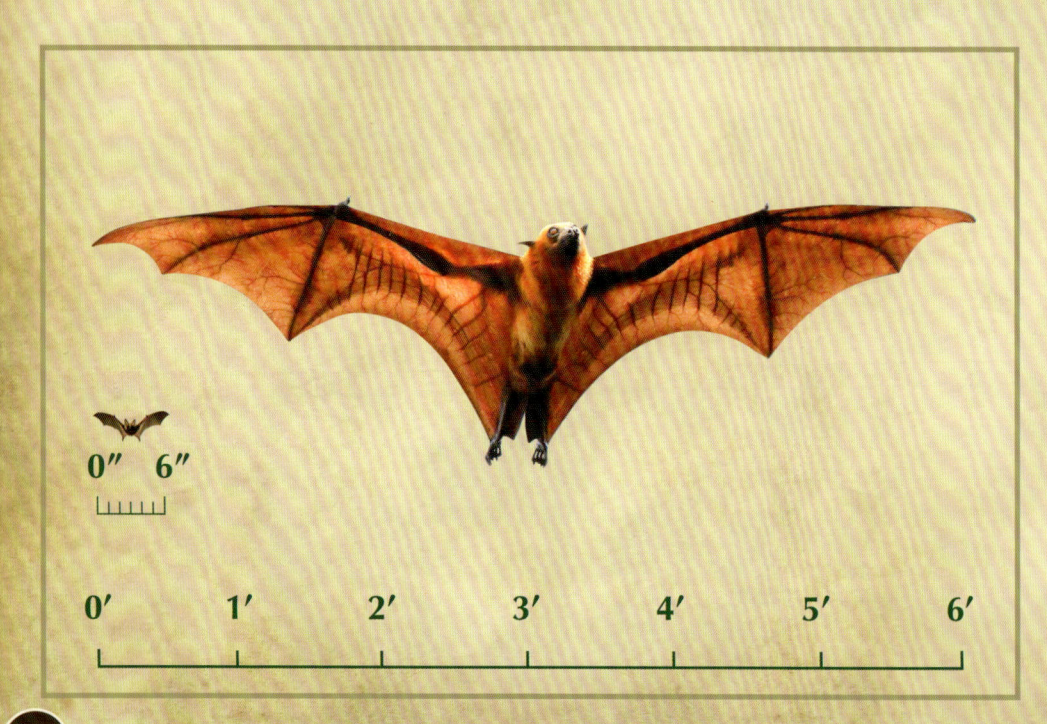

Bumblebee Bat

Flying Fox Bat

0" 6"

0'   1'   2'   3'   4'   5'   6'

Vampire bats are the only bats that drink blood. There are only a few species of vampire bats.

Vampire bats mostly drink blood from cows. They don't drink much. And the cows don't even feel it!

Other Bats

The fisherman bat swoops down and grabs fish out of the water.

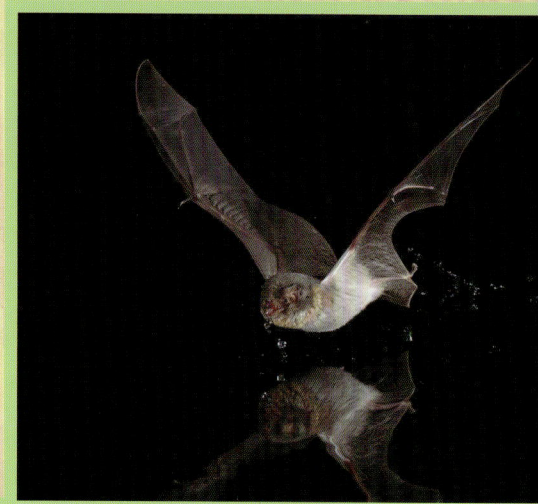

Stripe-faced vampire bats eat fruit, not blood.

The tent-making bat makes tents out of leaves.

Which Are Real Species of Bats?

Pink Bat	Squirrel Bat
Ghost Bat	Mouse-Tailed Bat
Horseshoe Bat	Pointy-Nosed Bat
Squeaky Bat	Spectacled Bat

1) Mouse-Tailed Bat, 2) Ghost Bat, 3) Horseshoe Bat, 4) Spectacled Bat

Pup's First Flight

Pup: A baby bat.

Mother bats usually have just one **pup**. Mother bats live in **nursery colonies** when their pups are born.

Nursery Colony: A big group of mother and baby bats.

Bats give birth hanging upside-down. The pup falls into a pouch made by the tail of the wing.

The pup will cling to its mother while she hunts. They may do this for a few days or weeks. Then, the mother will start leaving it behind.

Hundreds or thousands of bats will live together. But bats will have a few close friends. They'll spend the most time with these friends.

Hearing to See

The sound bats make for echolocation is high. Humans can't hear it.

Bats make other sounds, too. Some of those sounds can be heard by people.

Bats click, purr, and buzz.

Bats at Home

Bats like to **roost** in sheltered places. They may roost in caves, old barns, or tree hollows.

A few species of bats roost in trees in the open.

Roost: To settle down and rest.

NORTH AMERICA

ATLANTIC OCEAN

PACIFIC OCEAN

SOUTH AMERICA

SOUTH ATLANTIC OCEAN

Bats Live Most Everywhere

ARCTIC OCEAN

EUROPE

ASIA

PACIFIC OCEAN

AFRICA

INDIAN OCEAN

AUSTRALIA

ANTARCTICA

They don't live in the polar regions. They don't live in extreme deserts. They don't live on a few islands. But they live pretty much everywhere else.

A Voice That Carries

Which bugs can the bat find?

Other things use echolocation.

20

Up All Night

Nocturnal: Active at night.

Most bats are **nocturnal**. They probably hunt at night so birds aren't hunting at the same time.

Hibernate: Go into deep sleep in winter.

Most bats **hibernate**. Some bats **migrate**.

Migrate: Move to a different place when the season changes.

Keeping Bugs

Bats eat…

| MOSQUITOS | MOTHS |
| BANANAS | BLOOD |

Bats mostly eat bugs. They can eat thousands in one night! This makes them great pest control.

at Bay

Some bats eat fruit.

Vampire bats only drink blood. If they go more than two nights without blood, they will die. But vampire bats will share blood with each other by throwing it up.

A Night in the Life...

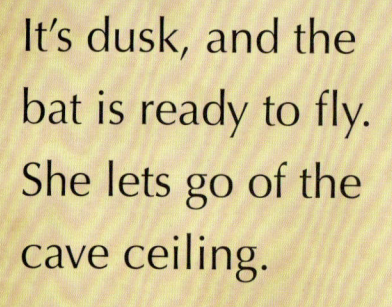

It's dusk, and the bat is ready to fly. She lets go of the cave ceiling.

She falls into flight with her friends.

They fly out of their cave to look for bugs.

# of a Little Brown Bat

She lets out a sound. The sound comes back and hits her ears. Now she knows where the mosquito is.

In just an hour, she eats 1,000 bugs. She rests before going out to hunt for more. She will hunt, eat, and rest all night.

Food Web

Bats eat bugs and fruit. Snakes, owls, and hawks eat bats.

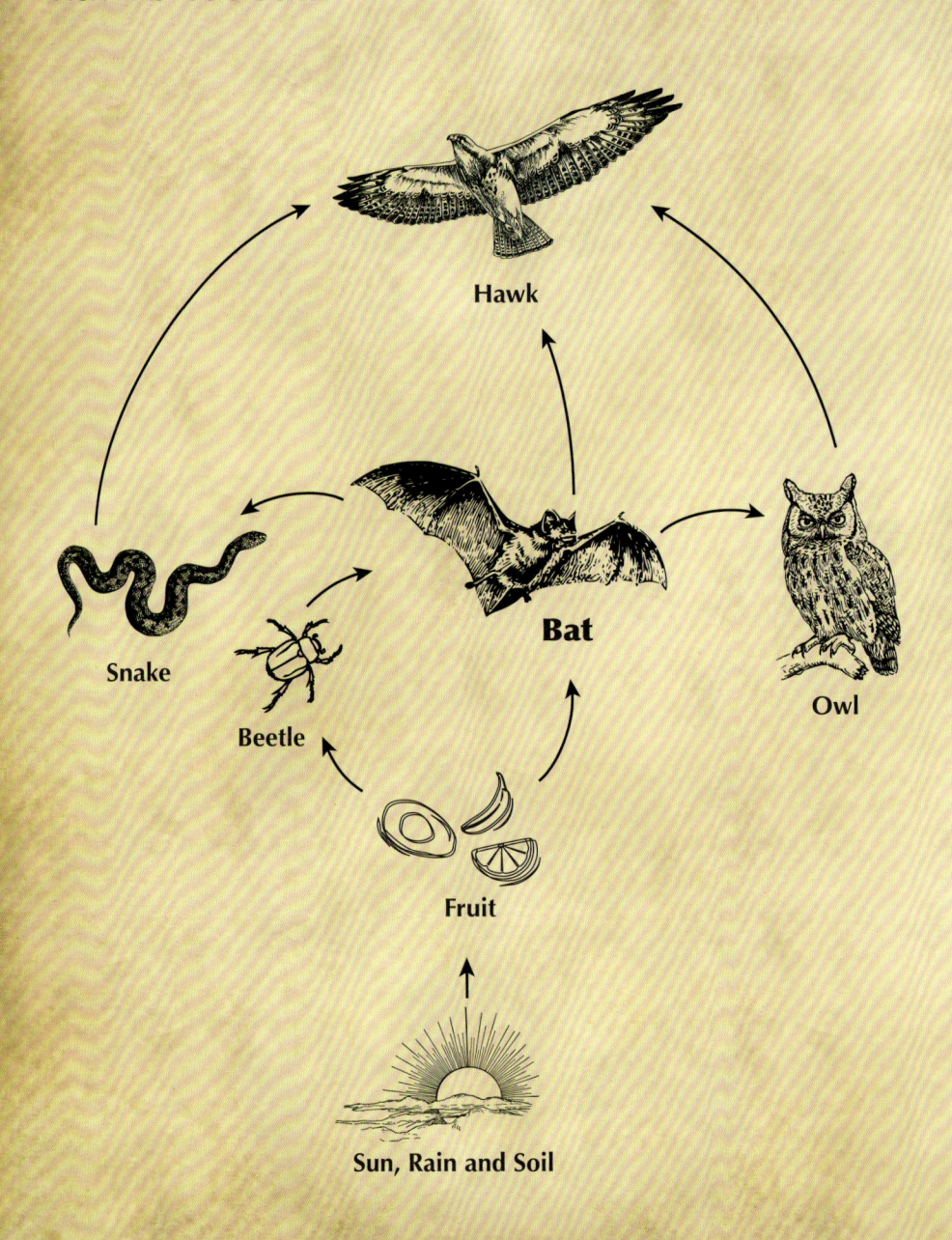

Hawk

Snake

Beetle

Bat

Owl

Fruit

Sun, Rain and Soil

26

Dangers

Sickness is one of the biggest dangers to bats. A fungus called white-nose syndrome has killed many bats. Scientists are still trying to learn about this fungus.

Many people think bats always have rabies. Actually, less than one percent of bats get rabies.

Hand Wings

Bats have extra big hands. Their hands make up their wings.

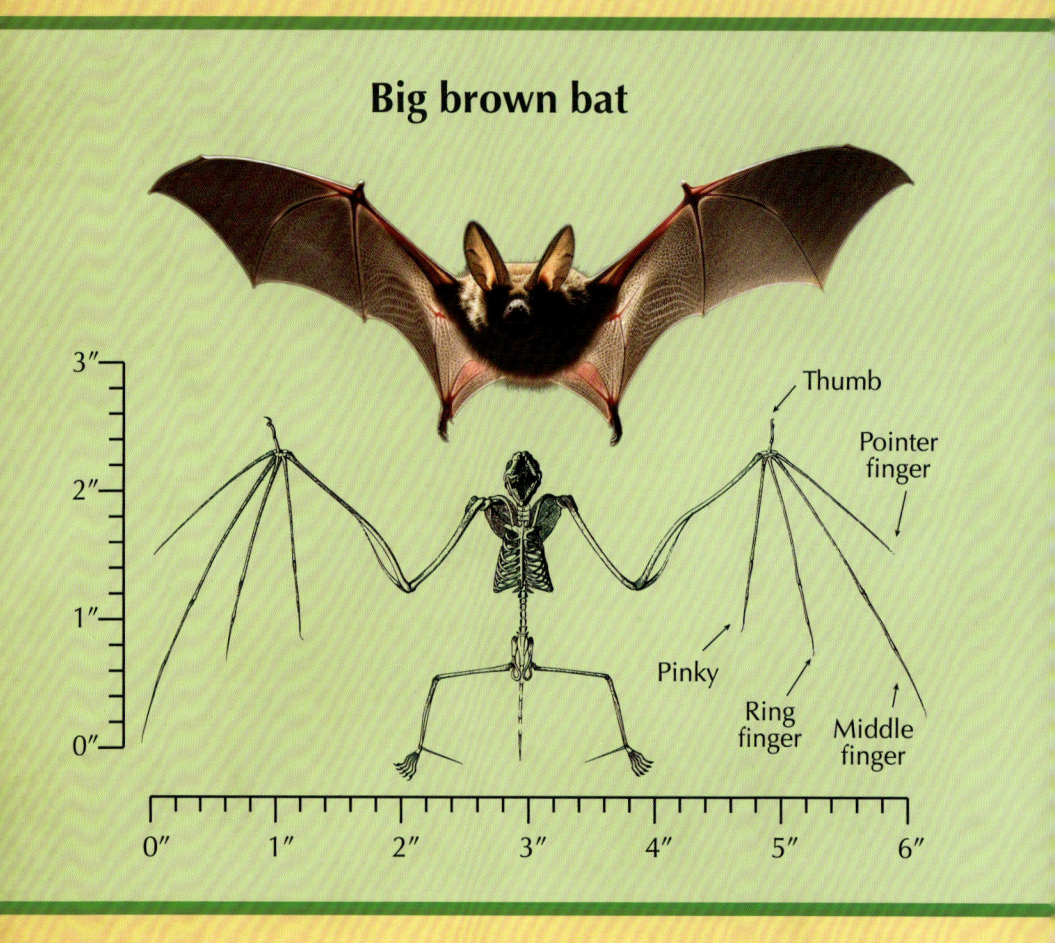

Big brown bat

Thumb
Pointer finger
Pinky
Ring finger
Middle finger

What if people had hands as big as a bat?

Have a grown-up help you measure yourself. Then, see how big your hands would be if you were a bat.

Grown-ups, to find out how long your child's hands and fingers would be, measure their height in inches. Multiply that number by 0.6. The number you get would be about the length of their third, fourth, and fifth digits if they were a big brown bat!

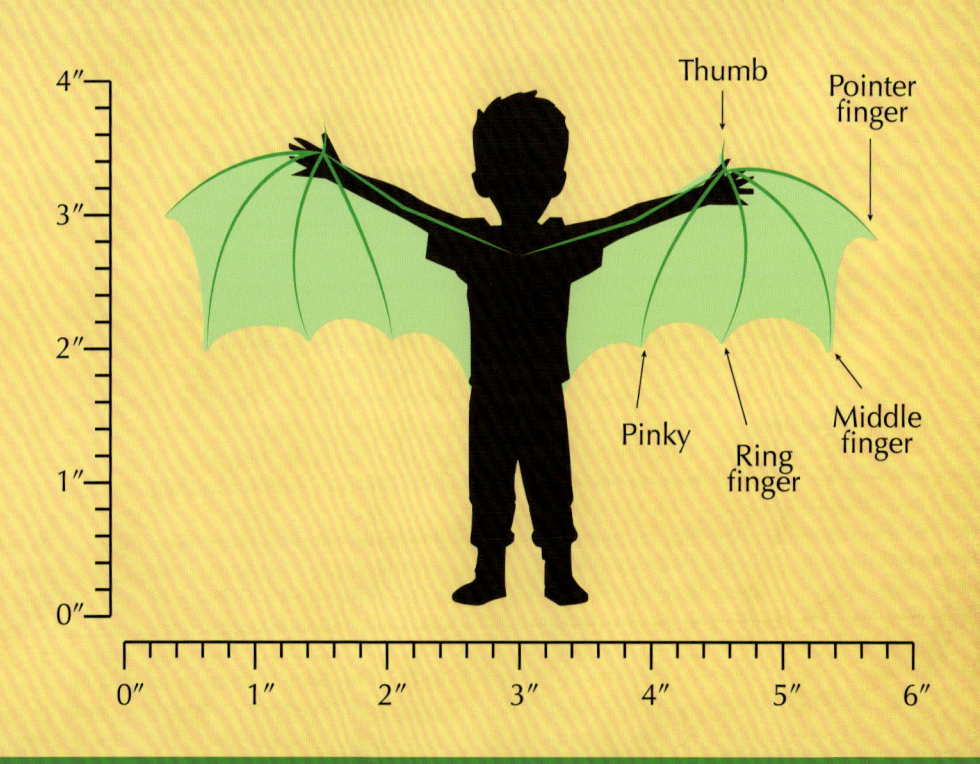

The second digit – the pointer finger – is shorter, a little over half that length. Bat thumbs are short! They're not part of the wing.

Glossary

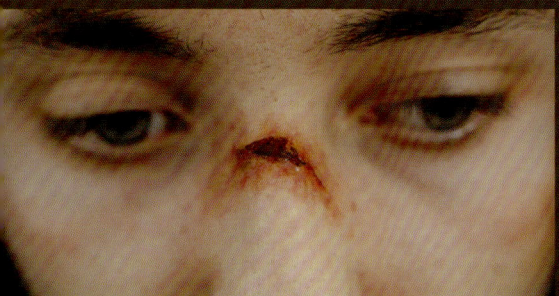

Clot: Blood that hardens into a clump.

Echolocation: To find out where things are by listening to an echo.

Guano: A build-up of poop from bats or seabirds.

Hibernate: Go into deep sleep in winter.

Mammal: Warm-blooded animals with hair or fur.

Migrate: Move to a different place when the season changes.

Nocturnal: Active at night.

Nursery Colony: A big group of mother and baby bats.

Pollen: Dust that helps plants form seeds.

Pollinator: An animal that moves pollen from one plant to another.

Pup: A baby bat.

Roost: To settle down and rest.

Funny Flyers

*The bat could see all right
Even late at night.
But when he used sound
To find things around
It was quite a sight.*

What's more amazing than a talking bat?

A spelling bee.

What do you get if you cross a computer with a vampire bat?

Love at first byte.

Why are baseball games at night?

Because bats sleep during the day.

Why do vampire bats drink blood?

Because coffee keeps them awake.

What's the first thing bats learn at school?

The alpha-bat.

How does a bat say hi to its mom?

With a sound wave.